The Cost of Convenience:

The Hidden Dangers of Automation

Simon Stanier

Table Of Contents

Chapter 1: Introduction

Automation is the process of using technology to perform tasks that were once carried out by humans. This can range from simple, repetitive tasks like data entry to more complex activities such as driving vehicles or performing surgeries. Automation has been an ongoing process for centuries, with advances in technology leading to new and innovative ways to automate various tasks. However, in recent years, the pace of automation has accelerated, and it has become increasingly prevalent in various industries.

One of the main drivers of this trend is the increasing availability and affordability of technology. With the rise of digital technologies like artificial intelligence, machine learning, and robotics, it has become easier and cheaper to automate various tasks. Companies across different sectors have been eager to adopt these technologies to improve their efficiency, productivity, and profitability.

The manufacturing industry was one of the first to embrace automation, with the development of assembly line technologies in the early 20th century. Since then, automation has spread to other sectors such as healthcare, transportation, finance, and retail. For example, in healthcare, automation has been used to diagnose diseases, monitor patients, and even perform surgeries. In finance, automation has been used to process transactions, analyze data, and manage risk. In retail, automation has been used to streamline supply chain operations, manage inventory, and personalize customer experiences.

Another factor contributing to the growth of automation is the increasing demand for speed and convenience. With customers expecting faster and more personalized service, companies are under pressure to deliver goods and services quickly and efficiently. Automation has become a key strategy for achieving this goal, as it can help companies to reduce costs, speed up processes, and improve accuracy.

Despite the benefits of automation, there are also concerns about its impact on the workforce. As more tasks are automated, there is a risk of job displacement, particularly for workers who perform routine or low-skilled tasks. This can lead to unemployment and economic inequality, as those who are unable to adapt to automation may struggle to find new employment.

Automation simply put is the process of using technology to perform tasks that were once carried out by humans. It has become increasingly prevalent in various industries due to advances in technology, increasing demand for speed and convenience, and the desire to improve efficiency and productivity. While automation offers many benefits, there are also concerns about its impact on the workforce and society as a whole. In the following chapters, we will explore these concerns in more detail and analyze the potential negative effects of automation.

While automation has many potential benefits, there are also several negative effects that can arise from its widespread adoption. In this chapter, we will preview some of the key negative effects of automation that will be discussed in more detail in subsequent chapters.

Job Displacement and Economic Inequality: One of the most significant negative effects of automation is the potential for job displacement. As more tasks become automated, there is a risk of job losses, particularly for workers who perform routine or low-skilled tasks. This can lead to economic inequality, as those who are unable to adapt to automation may struggle to find new employment.

Skills Gap and Retraining Challenges: As automation continues to evolve, it is creating new skill requirements that many workers may not possess. This can create a skills gap, where there is a mismatch between the skills that workers have and the skills that are required for new jobs. Retraining workers to acquire new skills can be challenging and costly, and not all workers may have access to the necessary training and education.

Psychological and Emotional Impact: Automation can also have negative psychological and emotional impacts on workers. When workers are replaced by machines, they may experience feelings of inadequacy, shame, and loss of purpose. This can lead to stress, depression, and other mental health issues, which can further exacerbate the negative effects of automation.

Loss of Human Interaction and Social Connections: As more tasks become automated, there is a risk of losing human interaction and social connections. This is particularly true in service industries, where automation can replace human-to-human interactions, leading to a loss of empathy and personalized service. This can also have negative impacts on mental health and social cohesion, as people become more isolated and disconnected from each other.

Environmental Impacts: Finally, automation can have negative environmental impacts, particularly in industries such as manufacturing and transportation. While automation can reduce waste and increase efficiency, it can also lead to increased energy consumption, pollution, and other environmental problems.

In the following chapters, we will delve deeper into these negative effects of automation, analyzing their causes, impacts, and potential solutions. While the negative effects of automation are significant, it is also important to remember that there are ways to mitigate these effects and create a more equitable and sustainable future.

Chapter 2: Job displacement and economic inequality

Automation has the potential to replace human workers in a variety of industries, leading to job displacement and unemployment. As machines become more sophisticated and capable, they can perform tasks that were once carried out by humans, often more efficiently and at a lower cost. While this can bring benefits to companies, it can also have negative impacts on workers and society as a whole.

One of the main ways that automation can lead to job displacement is by replacing routine and low-skilled tasks. These tasks are often simple and repetitive, and can be easily automated using robotics and artificial intelligence. For example, in manufacturing, assembly line tasks such as welding, painting, and packaging can be carried out by machines, reducing the need for human workers. In retail, self-checkout machines can replace human cashiers, and in transportation, autonomous vehicles can replace human drivers.

Another way that automation can lead to job displacement is by reducing the need for certain types of workers. For example, in healthcare, automation can reduce the need for certain types of medical technicians, such as radiology and laboratory technicians, by automating the diagnostic process. In finance, automation can reduce the need for certain types of clerical workers, such as data entry and processing.

While the benefits of automation are clear, job displacement can have negative impacts on workers and society as a whole. When workers lose their jobs due to automation, they may struggle to find new employment, particularly if they lack the necessary skills or education. This can lead to economic inequality and social unrest, as those who are unable to adapt to automation are left behind.

Furthermore, job displacement can have wider impacts on the economy. When workers lose their jobs, they have less disposable income to spend on goods and services, leading to a reduction in consumer demand. This can lead to a slowdown in economic growth, and in extreme cases, a recession.

To mitigate the negative impacts of job displacement, it is important to invest in retraining and education programs for workers. This can help them acquire the skills and knowledge necessary to adapt to the changing job market and find new employment opportunities. It is also important to create policies and programs that support workers who have been displaced by automation, such as unemployment benefits and job placement services.

In conclusion, automation has the potential to replace human workers in a variety of industries, leading to job displacement and unemployment. While the benefits of automation are significant, it is important to consider the negative impacts on workers and society as a whole, and to develop policies and programs to mitigate these impacts.

As automation continues to evolve, it is creating winners and losers in the job market, with those who have the skills and resources to adapt to automation benefiting, while those who do not may suffer. This can lead to economic inequality, where the benefits of automation are concentrated in the hands of a few, while the costs are borne by many.

One of the main ways that automation can create economic inequality is by increasing the demand for highly skilled workers. As automation replaces routine and low-skilled tasks, the demand for workers who can design, program, and maintain these systems is increasing. This has led to a growing skills gap, where there is a mismatch between the skills that workers have and the skills that are required for new jobs.

Highly skilled workers who are able to adapt to automation are in high demand and can command higher salaries and better working conditions. This can create a two-tiered job market, where a small group of highly skilled workers benefit from the benefits of automation, while the rest of the workforce is left behind. This can lead to a widening income gap and economic inequality.

Furthermore, the cost of adapting to automation can be high, particularly for workers who lack the necessary skills or education. Retraining and education programs can be expensive and time-consuming, and not all workers may have access to them. This can create a barrier to entry for certain types of jobs and exacerbate economic inequality.

Another way that automation can create economic inequality is by reducing the bargaining power of workers. When workers are replaced by machines, they may have less bargaining power and be more vulnerable to exploitation. This can lead to lower wages, worse working conditions, and less job security, particularly for low-skilled workers who are most at risk of being replaced by automation.

To mitigate the negative impacts of economic inequality, it is important to invest in education and training programs that can help workers acquire the skills and knowledge necessary to adapt to automation. It is also important to create policies and programs that support workers who are most at risk of being displaced by automation, such as unemployment benefits and job placement services. Additionally, policymakers can consider implementing policies that redistribute the benefits of automation, such as a universal basic income or progressive taxation.

In conclusion, automation has the potential to create economic inequality, as those who have the skills and resources to adapt to automation may benefit while those who do not may suffer. To mitigate these negative impacts, it is important to invest in education and training programs, create policies that support workers who are at risk of being displaced by automation, and consider implementing policies that redistribute the benefits of automation.

The increasing prevalence of automation has the potential to lead to widespread job displacement across many industries. While some argue that automation will create new jobs and opportunities, there are concerns about the potential long-term consequences of widespread job displacement. In this chapter, we will explore some of the potential consequences of job displacement due to automation.

One of the most immediate consequences of job displacement is the loss of income for workers. This can have significant economic impacts, particularly for low-income workers who may struggle to find new employment. Unemployment can also lead to social and psychological consequences, such as increased stress and anxiety, social isolation, and depression.

In the long term, job displacement due to automation can lead to a decrease in the demand for labor, which can have broader economic impacts. A decrease in demand for labor can lead to a decrease in wages, which can reduce consumer spending and slow economic growth. It can also lead to a decrease in the tax base, which can limit the government's ability to provide social programs and services.

Another potential consequence of widespread job displacement is the exacerbation of economic inequality. As we discussed in the previous chapter, automation can lead to a two-tiered job market, where a small group of highly skilled workers benefit from the benefits of automation, while the rest of the workforce is left behind. This can lead to a widening income gap and economic inequality, which can have significant social and political consequences.

Furthermore, widespread job displacement can lead to social unrest and political instability. When large numbers of people are unable to find employment, there is a risk of social unrest, protests, and even violence. This can lead to political instability, particularly in countries with weak democratic institutions or high levels of inequality.

To mitigate the potential long-term consequences of job displacement due to automation, it is important to invest in education and training programs that can help workers acquire the skills and knowledge necessary to adapt to automation. It is also important to create policies and programs that support workers who are most at risk of being displaced by automation, such as unemployment benefits, job placement services, and wage subsidies.

In addition, policymakers can consider implementing policies that redistribute the benefits of automation, such as a universal basic income or progressive taxation. These policies can help ensure that the benefits of automation are shared more equally and can help mitigate the negative long-term consequences of job displacement.

In conclusion, widespread job displacement due to automation has the potential to have significant long-term consequences, including economic impacts, exacerbation of economic inequality, and social and political instability. To mitigate these consequences, it is important to invest in education and training programs, create policies that support workers who are at risk of being displaced by automation, and consider implementing policies that redistribute the benefits of automation.

Chapter 3: Decreased human creativity and ingenuity

Automation has the potential to revolutionize many industries and improve efficiency and productivity. However, there is growing concern that automation may come at the cost of limiting human creativity and problem-solving ability. In this chapter, we will explore how automation can limit human creativity and problem-solving ability.

One of the ways that automation can limit human creativity is by reducing the need for creative problem-solving. When tasks are automated, the need for creative problem-solving is reduced, as the solution is often predetermined by the automation system. This can lead to a decrease in the development of problem-solving skills and creativity among workers, as they are less likely to be challenged to come up with new and innovative solutions.

Additionally, automation can lead to a reliance on standardized solutions rather than customized solutions. Automation systems are designed to follow predefined rules and procedures, which can limit the ability to develop customized solutions to complex problems. This can lead to a decrease in the development of innovative and creative solutions to complex problems, as the focus shifts to following standardized solutions.

Furthermore, automation can limit human creativity by reducing the ability to experiment and take risks. Automation systems are designed to operate within specific parameters and to minimize errors and deviations. As a result, there is less room for experimentation and risk-taking, which can limit the development of innovative and creative solutions to complex problems.

Another way that automation can limit human creativity is by reducing the opportunities for collaboration and teamwork. Automation systems can lead to a decrease in the need for human interaction and collaboration, as tasks can be completed by machines without human input. This can limit the ability to develop creative solutions to complex problems through collaboration and teamwork, which can result in missed opportunities for innovation and creativity.

To mitigate the potential limitations of automation on human creativity and problem-solving ability, it is important to focus on developing skills and competencies that are not easily automated, such as creativity, critical thinking, and emotional intelligence. Additionally, organizations can encourage experimentation, risk-taking, and collaboration, and provide opportunities for workers to engage in creative problem-solving.

Automation has the potential to limit human creativity and problem-solving ability by reducing the need for creative problem-solving, leading to a reliance on standardized solutions, limiting experimentation and risk-taking, and reducing opportunities for collaboration and teamwork. To mitigate these limitations, it is important to focus on developing skills and competencies that are not easily automated, encourage experimentation and risk-taking, and provide opportunities for workers to engage in creative problem-solving.

Automation has become a vital part of many industries, and its benefits are undeniable. However, the increasing reliance on automation for certain tasks can also have negative consequences, such as reducing innovation and originality. In this chapter, we will explore how relying on automation for certain tasks can reduce innovation and originality.

One of the ways that relying on automation can reduce innovation and originality is by creating a culture of conformity. Automation systems are designed to follow specific rules and procedures, which can create a culture that values conformity and adherence to predetermined standards. This can discourage workers from taking risks or proposing new ideas, as there is a fear that deviating from the established procedures could lead to errors or mistakes.

Additionally, relying on automation can limit the opportunity for workers to develop new skills and expertise. When tasks are automated, workers may be less likely to engage in creative problem-solving and critical thinking, which are essential skills for innovation and originality. This can lead to a loss of creativity and expertise, as workers are less likely to be challenged to develop new ideas and approaches.

Furthermore, automation can reduce the incentive for organizations to invest in research and development. When tasks are automated, organizations may be less likely to invest in research and development, as there is a perception that the automated system is already efficient and effective. This can limit the opportunities for organizations to develop new technologies and processes that can lead to innovation and originality.

Moreover, automation can limit the ability to adapt to changing circumstances. When tasks are automated, the system is designed to operate within specific parameters and may not be able to adapt to changes in the environment or new requirements. This can limit the ability to develop new and innovative solutions to emerging challenges, leading to a lack of innovation and originality.

To mitigate the potential negative impact of relying on automation for certain tasks, it is important to foster a culture of creativity and innovation. Organizations can encourage workers to take risks and propose new ideas, and provide opportunities for them to develop new skills and expertise. Additionally, organizations can invest in research and development to develop new technologies and processes that can lead to innovation and originality.

In summary relying on automation for certain tasks can reduce innovation and originality by creating a culture of conformity, limiting the opportunity for workers to develop new skills and expertise, reducing the incentive for organizations to invest in research and development, and limiting the ability to adapt to changing circumstances. To mitigate these limitations, it is important to foster a culture of creativity and innovation, encourage workers to take risks and propose new ideas, and invest in research and development.

While automation has brought many benefits to various industries, it has also hindered human creativity in some fields. In this chapter, we will explore some examples of industries or fields where automation has hindered human creativity.

Manufacturing Industry: The manufacturing industry has been one of the earliest adopters of automation. Assembly line robots have been used for decades to perform repetitive tasks, such as welding, painting, and packaging. However, the increasing reliance on automation has limited the creativity of human workers, who are often relegated to overseeing the machines rather than actively participating in the manufacturing process.

Music Industry: The music industry has also been impacted by automation. With the rise of digital audio workstations and virtual instruments, it has become easier than ever to create music without the need for live musicians. While this has made music production more accessible, it has also limited the opportunities for human musicians to showcase their creativity and expressiveness.

Graphic Design Industry: The graphic design industry has also seen significant automation in recent years. With the rise of artificial intelligence and machine learning, it has become possible to create designs without human intervention. However, this has limited the creative input of human designers, who are often reduced to selecting from a limited set of pre-designed templates.

Transportation Industry: The transportation industry has been heavily impacted by automation in recent years, with the introduction of autonomous vehicles and drones. While this has brought many benefits, such as increased safety and efficiency, it has also limited the opportunities for human drivers and pilots to showcase their creativity and problem-solving skills.

Journalism Industry: The journalism industry has also seen automation in recent years, with the rise of automated news writing and AI-powered content generation. While this has made it possible to produce news articles quickly and efficiently, it has also limited the creative input of human writers and reporters, who are often reduced to verifying facts and editing pre-written content.

Cooking Industry: The cooking industry has seen the rise of automated cooking machines that can perform tasks like chopping vegetables, mixing ingredients, and even cooking food. While these machines can save time and increase efficiency, they also limit the opportunities for chefs to showcase their creativity and culinary skills.

Architecture Industry: The architecture industry has seen automation in the form of computer-aided design (CAD) software, which can generate blueprints and 3D models of buildings quickly and accurately. While this technology can improve efficiency and accuracy in the design process, it can also limit the creative input of human architects who may rely too heavily on the software.

Healthcare Industry: The healthcare industry has seen automation in the form of robots that can perform surgeries and diagnose diseases. While these machines can increase accuracy and reduce the risk of human error, they also limit the opportunities for human doctors to showcase their problem-solving skills and adapt to unexpected complications during a procedure.

Retail Industry: The retail industry has seen the rise of automated checkout machines, which can process transactions and handle inventory management. While these machines can increase efficiency and reduce the need for human cashiers, they also limit the opportunities for human workers to showcase their customer service skills and build relationships with customers.

Education Industry: The education industry has seen automation in the form of online learning platforms and educational software, which can provide personalized learning experiences for students. While these technologies can improve access to education and help students learn at their own pace, they also limit the opportunities for human teachers to provide individualized instruction and foster creative problem-solving skills in their students.

These examples illustrate how automation can limit the opportunities for human workers to showcase their creativity and problem-solving abilities, and instead relegate them to oversee or manage the machines. It is important to consider the potential impact of automation on human creativity in various industries and find ways to strike a balance between the use of automation and the need for creative input from human workers.

Chapter 4: Increased reliance on technology and its consequences

As automation becomes increasingly prevalent in various industries, it is leading to a growing reliance on technology. While technology can offer many benefits, there are also potential negative effects of this increased reliance on technology that must be considered.

Dependence on technology: As automation takes over more tasks previously done by humans, people may become increasingly dependent on technology to complete these tasks. This dependence can lead to a loss of important skills and abilities, making it more difficult to adapt if the technology fails or if new tasks arise that cannot be automated.

Loss of privacy: Increased reliance on technology can also lead to a loss of privacy, as data is collected and stored by machines. This data can be used for targeted advertising, and can also be vulnerable to hacking and cyber attacks.

Health and safety concerns: The use of technology can also pose health and safety concerns, especially if the technology is not properly maintained or monitored. For example, if automated machines are not properly calibrated, they can pose physical risks to workers and the public.

Social isolation: As more tasks are automated, there is a risk that human interaction will be reduced. This can lead to social isolation and a breakdown in social connections, which can have negative impacts on mental health.

Job loss: As discussed in previous chapters, increased automation can lead to job displacement and loss of income for workers. This can have negative impacts on mental health, and can also lead to economic inequality and social unrest.

Overall, increased reliance on technology due to automation can have a range of negative effects on individuals and society as a whole. It is important to carefully consider the potential impacts of automation and find ways to mitigate any negative effects while still reaping the benefits of increased efficiency and productivity.

One potential negative consequence of automation is the over-reliance on technology. When humans delegate tasks to machines, they may start to rely too heavily on these automated systems. This can create potential problems when technology fails or malfunctions, leaving people unable to complete tasks or solve problems without the assistance of automated systems.

Limited human ability: When people rely too heavily on technology, they may begin to lose important skills and abilities that were once necessary to complete tasks. For example, if people rely on automated navigation systems to get around, they may start to lose the ability to navigate using a map or compass. This can create problems if the technology fails, leaving people without the necessary skills to complete the task at hand.

Increased risk of error: Automation can also create a false sense of security, leading people to trust the technology to perform tasks perfectly without human intervention. However, technology can fail or make mistakes, leading to errors and potential risks. When people rely too heavily on automated systems, they may not be able to identify or correct errors, leading to potentially dangerous situations.

Reduced problem-solving ability: When people rely too heavily on automated systems, they may also lose the ability to problem-solve and think creatively. This can create problems when technology fails or malfunctions, leaving people without the necessary problem-solving skills to overcome the issue.

Increased frustration and stress: When people rely too heavily on technology, they may become frustrated and stressed when it fails or malfunctions. This can lead to a loss of productivity, increased stress levels, and negative impacts on mental health.

Power outages: When automated systems rely on electricity, power outages can disrupt entire systems. For example, if a power outage occurs at a hospital that relies heavily on automated medical equipment, patients may be left without critical life-saving care.

Cyber attacks: Automated systems are vulnerable to cyber attacks, which can compromise security and lead to loss of data or disruption of critical systems. Over-reliance on technology can make organizations more vulnerable to these types of attacks, as they may not have sufficient human intervention in place to detect and mitigate potential threats.

Transportation breakdowns: When transportation systems rely heavily on automated technology, breakdowns or malfunctions can cause significant problems. For example, if a self-driving car malfunctions on a busy highway, it can create a dangerous situation for other drivers and passengers on the road.

Reduced customer service: When companies rely too heavily on automated customer service systems, it can lead to reduced quality of service and customer satisfaction. For example, if a customer is having trouble with an automated phone system, they may become frustrated and switch to a competitor with more personalized customer service.

Overall, the over-reliance on technology due to automation can create potential problems when technology fails or malfunctions. It is important for individuals and organizations to carefully consider the potential risks and limitations of automated systems, and to maintain a balance between technology and human intervention. This can help to ensure that people are prepared to handle problems when technology fails, and can maintain important skills and abilities even in a highly automated world.

As automation and technology continue to advance, there is a growing concern about the potential long-term consequences of increased dependence on these systems. While automation and technology can provide significant benefits in terms of efficiency, productivity, and convenience, there are also potential risks and drawbacks to consider. In this chapter, we will analyze some of the potential long-term consequences of increased dependence on automation and technology.

Reduced Human Skills and Creativity: As automation takes over many of the tasks that were once performed by humans, there is a risk that important skills and creativity will be lost. If people become too dependent on automation, they may lose the ability to perform tasks or solve problems without the assistance of machines. This could lead to a decline in important human skills such as critical thinking, problem-solving, and creativity.

Job Displacement: One of the most significant consequences of automation is the displacement of jobs. As machines become more capable of performing tasks that were once done by humans, many jobs are becoming obsolete. This can lead to unemployment and economic inequality, as those who do not have the necessary skills to work with machines are left behind.

Cybersecurity Risks: As automation and technology become more prevalent, there is a growing risk of cyber attacks and other security breaches. These attacks can compromise sensitive data and disrupt critical systems, leading to significant economic and social consequences.

Environmental Impact: Automation and technology have the potential to impact the environment in significant ways. For example, the manufacturing and disposal of electronic devices can lead to pollution and environmental degradation. Additionally, automation can lead to increased energy consumption, which can contribute to climate change.

Social Isolation: As people become more dependent on technology for communication and social interaction, there is a risk of increased social isolation. This can lead to a decline in social skills and an increased risk of mental health problems such as depression and anxiety.

Loss of Control: Finally, as automation and technology become more advanced, there is a risk that humans will lose control over these systems. This could lead to a loss of autonomy and independence, as machines become more capable of making decisions and taking actions without human intervention.

Overall, these potential long-term consequences of increased dependence on automation and technology highlight the need for careful consideration of the risks and benefits of these systems. While automation and technology can provide significant benefits, it is important to ensure that humans maintain control over these systems and that we do not become too dependent on them at the expense of important human skills and creativity.

Chapter 5: Ethical concerns with automation

As automation becomes more prevalent in various industries, there are increasing concerns about the ethical implications of these systems. Automation can raise a variety of ethical issues, including privacy, data security, and bias. In this chapter, we will discuss these ethical implications and their potential impact on society.

Privacy: Automation can involve the collection and use of large amounts of personal data. This raises concerns about privacy, as individuals may be uncomfortable with the idea of their personal information being collected and used without their consent. Additionally, the use of automated systems can lead to increased surveillance, which can infringe on individual privacy rights.

Data Security: The collection and use of personal data by automated systems can also raise concerns about data security. If this data is not properly secured, it can be vulnerable to cyber attacks and other security breaches. This can compromise the privacy of individuals and lead to significant economic and social consequences.

Bias: Automated systems can also be prone to bias, as they may be programmed by humans who hold certain biases and assumptions. This can lead to unfair treatment of certain individuals or groups, as automated systems may make decisions based on inaccurate or incomplete data.

Lack of Accountability: Automated systems can also raise concerns about accountability. If something goes wrong with an automated system, it may be difficult to determine who is responsible for the error or who should be held accountable. This can create a lack of transparency and trust in the system.

Dependence on Technology: Finally, automation can raise concerns about dependence on technology. If people become too dependent on automated systems, they may lose the ability to perform tasks or make decisions without the assistance of machines. This can lead to a loss of autonomy and independence, as well as a potential loss of important human skills and creativity.

Job Displacement: As we discussed in earlier chapters, automation can lead to job displacement and unemployment. This raises ethical questions about the responsibility of companies and governments to ensure that workers are able to adapt to the changing job market and that they are provided with support during the transition.

Algorithmic Bias: Automated decision-making systems often use algorithms that are trained on historical data. If this data is biased in any way, it can lead to algorithmic bias, which can have negative consequences for certain groups of people. For example, if a facial recognition system is trained on data that is primarily made up of images of white people, it may not be as accurate at recognizing faces of people of colour.

Human Oversight: Automated systems can make decisions that have a significant impact on people's lives. It is important to ensure that there is human oversight of these systems to ensure that they are making fair and ethical decisions. This oversight can include the ability to review and challenge decisions made by automated systems.

Transparency: The decision-making processes used by automated systems can be complex and difficult to understand. This raises concerns about transparency, as individuals may not be able to understand how decisions are being made or why they are being made. It is important to ensure that automated systems are transparent and that people are able to understand how decisions are being made.

Security: Automated systems can also raise concerns about security. For example, if an autonomous vehicle is hacked, it could pose a threat to the safety of passengers and other people on the road. It is important to ensure that automated systems are designed with security in mind and that they are regularly tested and updated to address potential vulnerabilities.

Environmental Impact: Automated systems can also have a significant impact on the environment. For example, if automated factories increase production, this could lead to increased pollution and other negative environmental impacts. It is important to consider the environmental impact of automation and to ensure that automated systems are designed to minimize their environmental impact.

International Relations: As automation becomes more prevalent, it can have an impact on international relations. For example, if automated factories in one country lead to job displacement in another country, this could lead to tensions between the two countries. It is important to consider the international implications of automation and to ensure that it is implemented in a way that promotes global cooperation and understanding.

Overall, the ethical implications of automation highlight the need for careful consideration of the risks and benefits of these systems. It is important to ensure that automated systems are designed and implemented in a way that respects privacy, promotes data security, and minimizes bias. Additionally, it is important to ensure that these systems are transparent and accountable, and that humans maintain control over the decision-making process. By addressing these ethical implications, we can ensure that automation is used in a responsible and ethical manner that benefits society as a whole.

Automation has the potential to revolutionize many industries and improve efficiency, but it also comes with risks. One major risk associated with automation is the possibility of autonomous systems malfunctioning or being used for harmful purposes. In this chapter, we will examine these risks in more detail.

One risk of automation is the potential for autonomous systems to malfunction. As more tasks are automated and machines become more autonomous, there is a greater risk of malfunctions occurring. If an autonomous system malfunctions, it can cause serious damage or injury. For example, if an autonomous vehicle malfunctions while driving, it could cause an accident.

Another risk of automation is the possibility of autonomous systems being used for harmful purposes. For example, drones equipped with weapons could be used for military purposes or to carry out terrorist attacks. Autonomous systems could also be used to hack into computer systems or carry out cyberattacks. The use of autonomous systems for harmful purposes raises ethical concerns and highlights the need for strict regulations and oversight.

Additionally, automation could potentially lead to job displacement, as we discussed earlier in this book. This could lead to economic instability and social unrest. Furthermore, if certain industries or regions are heavily reliant on automation, they could be vulnerable to disruptions if the systems fail or if there is a lack of maintenance or repair.

Finally, the rapid pace of technological change and automation means that regulations and laws may not be able to keep up. This could lead to a lack of oversight and accountability, leaving the potential for unscrupulous actors to take advantage of the technology for their own purposes.

Overall, while automation has the potential to bring many benefits, it is important to be aware of the potential risks and to take steps to mitigate them. This includes developing regulations and oversight to ensure that autonomous systems are used safely and ethically, investing in maintenance and repair to prevent malfunctions, and ensuring that the benefits of automation are shared fairly across society.

As automation becomes more prevalent in various industries, organizations and individuals must consider the ethical implications of using this technology. In this chapter, we will examine the ethical responsibility of organizations and individuals to ensure that automation is used ethically.

Organizations have a responsibility to ensure that automation is used in a way that is consistent with ethical principles. This includes ensuring that the technology is safe and reliable, and that it does not cause harm to individuals or society. Organizations must also consider the impact of automation on the workforce, and take steps to mitigate any negative effects on workers. This may include providing retraining or other forms of support to workers whose jobs are being automated.

Individuals also have a responsibility to ensure that automation is used ethically. This may include raising concerns if they believe that the technology is being used in a way that is unethical or harmful. Individuals must also consider their own use of automation, and ensure that they are using it in a way that is consistent with ethical principles.

One key ethical consideration when it comes to automation is privacy. As more tasks are automated, there is a greater risk of personal data being collected and used without consent. Organizations and individuals must take steps to ensure that personal data is collected and used in a way that is consistent with ethical principles, such as obtaining consent and ensuring that data is protected.

Another ethical consideration is the potential for automation to perpetuate bias and discrimination. If the algorithms that are used in automated systems are biased, they can perpetuate existing social inequalities. Organizations must take steps to ensure that their algorithms are fair and unbiased, and that they do not perpetuate discrimination.

Finally, organizations and individuals have a responsibility to ensure that automation is not used for harmful purposes. This includes ensuring that autonomous systems are not used for military purposes or to carry out terrorist attacks, and that they are not used to harm individuals or society.

In conclusion, organizations and individuals must consider the ethical implications of using automation, and take steps to ensure that the technology is used in a way that is consistent with ethical principles. This includes ensuring that automation is safe, reliable, and does not cause harm, protecting personal data, avoiding bias and discrimination, and ensuring that the technology is not used for harmful purposes. By taking a responsible and ethical approach to automation, we can ensure that this technology brings benefits to society while minimizing the potential risks and negative consequences.

Chapter 6: Social and psychological impacts of automation

As automation becomes more prevalent in various industries, there are potential social and psychological impacts that must be considered. In this chapter, we will discuss the potential social and psychological impacts of automation, including the loss of human interaction and social connections due to automation.

One of the most significant social impacts of automation is the potential loss of jobs and the resulting social and economic consequences. As we have discussed in previous chapters, automation has the potential to displace large numbers of workers, particularly those in industries such as manufacturing and transportation. This can lead to job loss, economic insecurity, and social isolation.

In addition to the economic impacts, automation can also have psychological impacts on individuals. For example, automation can lead to a loss of human interaction and social connections. In industries where automation is prevalent, workers may find themselves working alongside machines rather than other human beings. This can lead to feelings of loneliness and isolation, which can have negative impacts on mental health.

Another potential psychological impact of automation is the feeling of loss of control. As automation takes over tasks that were once performed by humans, individuals may feel that they are losing control over their work and their lives. This can lead to feelings of powerlessness and frustration, which can also have negative impacts on mental health.

Moreover, the implementation of automation can also lead to social inequality. Those who have the necessary skills and resources to adapt to automation may benefit, while those who do not may suffer. This can exacerbate existing social inequalities and lead to feelings of resentment and mistrust between different groups in society.

Finally, the use of automation may also lead to ethical concerns related to the use of personal data, privacy, and the potential for discrimination. If automation is not used in a way that is consistent with ethical principles, it can lead to further erosion of trust in institutions and exacerbate social divisions.

In conclusion, there are potential social and psychological impacts of automation that must be considered. The loss of human interaction and social connections due to automation can lead to feelings of loneliness and isolation, while the feeling of loss of control can lead to frustration and feelings of powerlessness. Moreover, the potential for social inequality and ethical concerns related to the use of personal data and privacy must also be considered. By understanding these potential impacts, we can work to mitigate the negative consequences of automation and ensure that it is used in a way that benefits society as a whole.

As automation becomes more prevalent in various industries, there is growing concern about its impact on job satisfaction and stress levels for workers. In this chapter, we will analyze how automation can lead to decreased job satisfaction and increased stress levels for workers.

One of the main ways in which automation can lead to decreased job satisfaction is through the loss of meaningful work. When workers are replaced by machines, they may find that their jobs become more repetitive and less engaging, leading to a sense of boredom and disengagement. Moreover, workers may feel that their skills and expertise are no longer valued, leading to a loss of self-esteem and motivation.

Another way in which automation can lead to decreased job satisfaction is through the loss of autonomy and control over one's work. When tasks are automated, workers may feel that they have little control over the process and may be forced to follow rigid procedures and protocols. This can lead to a sense of powerlessness and frustration, which can erode job satisfaction over time.

Moreover, automation can also increase stress levels for workers. For example, when workers are replaced by machines, they may fear for their job security and feel anxious about their future. This can lead to increased stress levels, which can have negative impacts on both physical and mental health. Moreover, workers may feel pressure to work at a faster pace or to meet more demanding performance targets, which can lead to burnout and fatigue.

Finally, the implementation of automation can also lead to social and cultural changes that can impact job satisfaction and stress levels. For example, when workers are replaced by machines, the social dynamics of the workplace may change, leading to feelings of isolation and alienation. Moreover, workers may feel that they are no longer part of a larger community or culture, leading to a loss of identity and purpose.

Chapter 7: Case studies and examples

There are numerous examples of industries and companies where automation has had negative effects, ranging from job displacement and economic inequality to decreased job satisfaction and increased stress levels for workers. In this chapter, we will provide specific case studies and examples of industries or companies where automation has had negative effects.

Amazon

Amazon has been a pioneer in automation, using robots and automated systems to sort and ship products in its fulfilment centres. However, this automation has also had negative effects on workers. For example, workers have reported increased stress levels due to the pace of work and the pressure to meet demanding performance targets. Moreover, the use of robots has also led to job displacement, with workers being replaced by machines.

Self-driving trucks

The rise of self-driving trucks has raised concerns about the impact on truck drivers. According to a report by the International Transport Forum, self-driving trucks could replace as many as 2.2 million truck drivers in the United States alone. This could have significant negative effects on the job market and could lead to economic inequality.

Manufacturing industry

The manufacturing industry has been one of the most heavily impacted by automation. Robots and automated systems have replaced workers in many tasks, leading to job displacement and economic inequality. Moreover, the loss of meaningful work has also led to decreased job satisfaction and increased stress levels for workers.

Call centres

Call centres have also been impacted by automation, with many tasks being automated using voice recognition and other technologies. This has led to job displacement and decreased job satisfaction for workers, who may find their work becoming more repetitive and less engaging.

Retail industry

The retail industry has also been impacted by automation, with self-checkout machines and other automated systems replacing workers in some tasks. This has led to job displacement and economic inequality, as well as decreased job satisfaction for workers.

Healthcare industry

The healthcare industry has also seen the rise of automation, with robots and other technologies being used to perform tasks such as surgery and diagnostics. While these technologies can improve patient outcomes, they can also lead to job displacement and decreased job satisfaction for healthcare workers.

Retail industry

With the rise of e-commerce and automation in warehousing and logistics, traditional brick-and-mortar retailers have struggled to compete. As a result, many retail workers have lost their jobs or have been forced to accept lower-paying positions in the gig economy.

Journalism industry

The rise of automated news writing software has led to concerns about the quality and accuracy of news reporting. Critics argue that automated news writing lacks the nuance and context of human-written stories, and may also perpetuate bias and misinformation.

The negative effects of automation on various industries and workers are not the result of a single factor, but rather a complex interplay between government policies, corporate decisions, and technological developments. In this chapter, we will analyze some of the key factors that have contributed to the negative effects of automation.

Government Policies:

Lack of regulation: One of the main factors that have contributed to the negative effects of automation is the lack of government regulations on the use of automation in various industries. The absence of clear policies and guidelines has allowed companies to prioritize profit over worker rights and safety, leading to increased job insecurity and inequality.

Insufficient retraining programs: As automation continues to displace workers, it is important for governments to provide sufficient retraining programs to help affected workers develop new skills and transition into new jobs. However, many governments have failed to provide adequate support for retraining, leaving workers with few options and exacerbating the negative effects of automation.

Corporate Decisions:

Focus on profit: The primary goal of many corporations is to maximize profits, often at the expense of workers and the broader society. In some cases, companies have implemented automation technologies solely to reduce labour costs, leading to job losses and a widening income gap.

Lack of transparency: Many companies have implemented automation technologies without sufficient transparency or input from workers, leading to a lack of trust and resentment among employees. This can further reduce job satisfaction and increase stress levels among workers.

Technological Developments:

Rapid technological advancements: The rapid pace of technological advancements has made it difficult for workers to keep up with the skills required to remain relevant in the job market. This skills gap has led to job losses and increased economic inequality, as workers who lack the necessary skills to adapt to automation are left behind.

Bias in AI systems: The increasing use of artificial intelligence (AI) and machine learning algorithms in automation has raised concerns about bias and discrimination. If not properly designed and implemented, these systems can perpetuate existing biases and lead to unfair treatment of certain groups.

In conclusion, the negative effects of automation on workers and society are the result of a complex interplay between government policies, corporate decisions, and technological developments. It is important for all stakeholders to work together to ensure that the benefits of automation are shared equitably, and that workers are not left behind in the process. This requires a combination of government regulations, corporate responsibility, and technological advancements that prioritize human welfare and well-being.

Chapter 8: The future of automation and its potential impact

Automation has already made significant inroads in various industries, ranging from manufacturing to healthcare, and continues to grow at a rapid pace. According to a report by the World Economic Forum, automation and artificial intelligence are expected to displace around 75 million jobs by 2022, while creating 133 million new jobs.

As automation becomes more prevalent, it is essential to examine the trends that are driving it and predict their potential impact on various industries and sectors.

Artificial Intelligence (AI) and Machine Learning

One of the primary drivers of automation is the rapid advancement of AI and machine learning. With the ability to learn from data and improve performance over time, these technologies are enabling machines to take on more complex tasks previously done by humans. For example, AI-powered chatbots are becoming increasingly common in customer service, while machine learning algorithms are being used to optimize supply chain management.

The impact of AI and machine learning is likely to be widespread, with some estimates suggesting that these technologies could automate up to 45% of all tasks done by workers in various industries.

Robotics

Robotics is another critical driver of automation, particularly in manufacturing and logistics. The use of robots in manufacturing has been increasing for decades, but recent advancements in technology have made them more versatile and efficient. For example, collaborative robots or cobots are now being used to work alongside human workers in various tasks, such as assembly and material handling.

The impact of robotics on various industries is likely to be significant, with some predicting that robots could replace up to 20 million manufacturing jobs worldwide by 2030.

Autonomous Vehicles

The development of autonomous vehicles, particularly in the transportation and logistics industries, is also driving automation. Self-driving trucks, for example, could potentially disrupt the entire trucking industry, replacing human drivers with machines that can operate 24/7.

The impact of autonomous vehicles on various industries is likely to be significant, with some predicting that they could automate up to 80% of all driving tasks by 2040.

Internet of Things (IoT)

The Internet of Things (IoT) is another trend that is driving automation, particularly in the industrial sector. With the ability to connect machines and devices, IoT is enabling automation in areas such as predictive maintenance, quality control, and inventory management.

The impact of IoT on various industries is likely to be significant, with some predicting that the global IoT market could reach $1.5 trillion by 2027.

Blockchain Technology

Blockchain technology, primarily known for its use in cryptocurrencies, is also being explored for its potential to automate various industries. For example, supply chain management can benefit from blockchain technology, as it can provide transparency, reduce fraud, and streamline processes.

The impact of blockchain technology on various industries is still uncertain, but some estimates suggest that the global blockchain market could reach $39.7 billion by 2025.

Automation is a growing trend that is transforming various industries, and its impact is likely to be significant in the coming years. The adoption of AI and machine learning, robotics, autonomous vehicles, IoT, and blockchain technology are likely to automate many tasks and disrupt industries, leading to significant job displacement and potential social and psychological impacts. However, automation also has the potential to create new opportunities and jobs, particularly in the fields of technology and engineering. As we continue to explore the potential of automation, it is essential to consider its impact on individuals, communities, and the economy and ensure that it is implemented ethically and responsibly.

As the prevalence of automation continues to grow, it is important for organizations and governments to consider the potential consequences and take steps to prepare for them. In this chapter, we will explore some recommendations for how organizations and governments can prepare for and respond to the potential consequences of automation.

Invest in retraining and upskilling programs

One of the most significant consequences of automation is the potential for job displacement. To mitigate this, organizations and governments should invest in retraining and upskilling programs to help workers transition to new roles. These programs can provide workers with the skills and knowledge necessary to work alongside automated systems or in new industries altogether.

Implement regulations to ensure ethical and responsible use of automation

As discussed earlier in the book, there are ethical implications associated with automation, including issues related to privacy, data security, and bias. Organizations and governments should implement regulations to ensure that automation is used ethically and responsibly. This may include measures to prevent bias in automated decision-making systems and to protect the privacy and security of personal data.

Foster a culture of innovation and creativity

While automation can limit human creativity and problem-solving ability, there are still many areas where human ingenuity is essential. To prepare for the potential consequences of automation, organizations and governments should foster a culture of innovation and creativity. This can include investing in research and development and creating incentives for employees to come up with new ideas.

Develop social safety nets

For those who are unable to adapt to the changing labour market, social safety nets can provide a crucial lifeline. Governments should consider implementing measures such as universal basic income or expanded unemployment benefits to help support workers who are displaced by automation.

Encourage collaboration between humans and machines

Rather than viewing automation as a threat to human workers, organizations and governments should encourage collaboration between humans and machines. This can involve designing systems that augment human capabilities rather than replace them, as well as providing training and support for workers to work alongside automated systems.

Embrace a growth mindset

Finally, organizations and governments should adopt a growth mindset when it comes to automation. This means viewing automation as an opportunity for growth and innovation rather than a threat to jobs and industry. By embracing a growth mindset, organizations and governments can work together to create a future where humans and machines work together to achieve shared goals.

The consequences of automation are complex and multifaceted, and there is no one-size-fits-all solution. However, by investing in retraining programs, implementing ethical regulations, fostering a culture of innovation, developing social safety nets, encouraging collaboration, and embracing a growth mindset, organizations and governments can prepare for and respond to the potential consequences of automation in a responsible and proactive manner. Ultimately, this will help to create a future where automation is used to benefit society as a whole, rather than a select few.

Chapter 9: Alternatives to automation

As we have seen in the previous chapters, the negative effects of automation on the workforce, economy, and society are significant. However, it is important to note that not all forms of automation are created equal, and there are alternative approaches that can mitigate some of these negative consequences. In this chapter, we will explore some of these alternative approaches, including the use of human labour and more traditional manufacturing methods.

One of the main arguments in favor of automation is its efficiency and speed. Machines can perform tasks faster and more accurately than humans, and they can do so around the clock without the need for breaks or rest. However, this argument overlooks the fact that human labor has unique advantages over machines, such as creativity, problem-solving ability, and adaptability. By harnessing these advantages, organizations can achieve a more balanced approach to automation that benefits both workers and the bottom line.

One approach to achieving this balance is to use human labor in conjunction with machines. This can involve a range of strategies, such as reorganizing work processes to allow for greater worker input and decision-making, providing training and education to help workers develop new skills and adapt to changing job requirements, and creating more flexible work arrangements that allow workers to work part-time or on a temporary basis. In some cases, it may also be possible to design machines that work alongside humans, rather than replacing them entirely.

Another alternative approach is to focus on more traditional manufacturing methods. This can include techniques such as craft production, which involves skilled workers creating products by hand using traditional tools and methods. While this approach may be less efficient than mass production using machines, it has several advantages, including greater customization, higher quality, and a lower environmental impact. It also provides opportunities for workers to develop and showcase their skills, leading to greater job satisfaction and a sense of pride in their work.

A third alternative approach is to focus on creating more sustainable and equitable models of production. This can involve a range of strategies, such as implementing fair labor practices, sourcing materials and components from sustainable sources, and designing products that are recyclable and biodegradable. By prioritizing sustainability and social responsibility, organizations can create a more ethical and equitable approach to automation that benefits both workers and the environment.

Of course, there are challenges to implementing these alternative approaches. For example, reorganizing work processes to allow for greater worker input and decision-making may require significant changes to organizational structures and cultures, and may require investment in training and development programs. Similarly, implementing more sustainable and equitable models of production may require a significant shift in organizational priorities and business practices. However, these challenges are not insurmountable, and the potential benefits of these alternative approaches make them worthy of consideration.

In conclusion, while automation has many benefits, its negative consequences cannot be ignored. However, by taking a more balanced and nuanced approach to automation, organizations can achieve the benefits of automation while minimizing its negative impacts on workers, society, and the environment. This may involve using human labor in conjunction with machines, focusing on more traditional manufacturing methods, and creating more sustainable and equitable models of production. By doing so, we can create a future that benefits everyone.

As we have discussed earlier, automation has brought about many benefits to various industries, such as increased efficiency, productivity, and cost savings. However, it is also crucial to consider the potential negative effects of automation, including job displacement, economic inequality, and loss of creativity and human interaction. As a result, it is essential to explore alternative approaches to automation and when and where they may be preferable.

One alternative approach to automation is the use of human labor. While this may not be as efficient as automation in terms of speed and consistency, it does offer several advantages. For example, using human labor can lead to increased job satisfaction and a sense of purpose for workers. Additionally, it can also promote social connections and foster a sense of community, which may be lost in automated settings. Moreover, human labor can be more adaptable to changes and can be more responsive to unexpected situations, such as changes in demand or product design.

Another alternative approach is the use of more traditional manufacturing methods. While these methods may be slower and less efficient than automation, they offer several advantages, such as increased job security and greater flexibility. For example, traditional manufacturing methods can be better suited for small-scale production runs or for producing customized products. Additionally, these methods often require a higher level of skill and craftsmanship, which can lead to higher job satisfaction and a greater sense of accomplishment for workers.

When deciding whether to use automation, human labor, or traditional manufacturing methods, it is important to consider several factors, such as the type of industry, the size of the company, the level of technological advancement, and the specific needs of the organization. For example, a small-scale manufacturer may find it more cost-effective to use human labor and traditional manufacturing methods rather than investing in expensive automation equipment. Conversely, a large-scale manufacturer may find that automation is necessary to meet the high demand for their products and to remain competitive in the market.

In general, it is recommended that organizations consider a balanced approach to automation, using it in conjunction with human labor and traditional manufacturing methods where appropriate. This approach can help to maximize the benefits of automation while minimizing the potential negative consequences. Additionally, it is crucial for organizations to invest in training programs and education for workers to ensure that they are equipped with the skills and knowledge necessary to adapt to new technologies and remain competitive in the job market.

Chapter 10: Overcoming the negative effects of automation

While automation can bring many benefits, it also poses significant risks and challenges, such as job displacement, economic inequality, and loss of human creativity. To mitigate these negative effects, it is essential to adopt strategies that help individuals and organizations adapt to the changing landscape of work. In this chapter, we will discuss some of the key strategies for overcoming the negative effects of automation.

Retraining and Reskilling Workers: As automation displaces human workers, it is crucial to retrain them for new jobs that require different skills. This requires investing in training programs that are tailored to the needs of individuals and industries. Governments and businesses can collaborate to provide such programs, which can include apprenticeships, on-the-job training, and vocational education. The goal is to help workers acquire new skills that are in demand and prepare them for the jobs of the future.

Implementing Ethical Guidelines: Automation can raise ethical concerns, such as data privacy, bias, and safety risks. To address these issues, it is essential to implement ethical guidelines that ensure the responsible use of automation. This includes ensuring that automated systems are transparent, fair, and unbiased. Companies must prioritize ethical considerations when designing and deploying automated systems. This can involve working with experts in ethics, data privacy, and cybersecurity.

Investing in Social Programs and Safety Nets: Automation can exacerbate economic inequality and put some individuals at risk of falling behind. To address this, it is essential to invest in social programs and safety nets that support those who are adversely affected by automation. This can include unemployment benefits, job training programs, and basic income programs. Governments and businesses must work together to ensure that these programs are well-funded, accessible, and effective.

Promoting Human Creativity: While automation can improve productivity, it can also limit human creativity and problem-solving ability. To overcome this, it is important to prioritize human creativity and ingenuity.

This means investing in education programs that emphasize critical thinking, creativity, and innovation. It also means encouraging collaboration and teamwork, which can help workers learn from each other and develop new ideas.

Balancing Automation with Human Labor: While automation can improve efficiency and productivity, it is important to strike a balance between automation and human labour. This means identifying tasks that are best suited for automation and those that require human skills, such as empathy, creativity, and judgment. It also means recognizing the value of human labor and creating jobs that leverage these skills.

As automation continues to become more prevalent in various industries, it is essential to consider the potential negative effects and strategies for mitigating them. Retraining workers, implementing ethical guidelines, and investing in social programs and safety nets are commonly proposed solutions.

Chapter 11: Trade union view on automation

Trade unions have traditionally been critical of automation, as they see it as a threat to job security and the livelihoods of their members. In many cases, trade unions have resisted the introduction of new technologies and automation, arguing that it will lead to job losses and lower wages for workers.

However, in recent years, some trade unions have begun to take a more nuanced view of automation, recognizing that it has the potential to improve working conditions and increase productivity if implemented correctly. Many trade unions are now advocating for a "just transition" to a more automated economy, which involves ensuring that workers are adequately trained and supported as new technologies are introduced.

Some trade unions have also called for a greater focus on social protections and safety nets, such as universal basic income, to ensure that workers are not left behind as automation continues to transform the economy.

Overall, while trade unions remain cautious about the potential negative effects of automation, they are increasingly recognizing that it is an inevitable part of the future of work. As a result, they are beginning to engage with employers and policymakers to find ways to ensure that automation is implemented in a way that benefits workers and society as a whole.

That being said Trade union have started to bring in automation agreements with employers. These are negotiated agreements between labour unions and management that address the impact of automation on the workforce. These agreements typically cover issues such as job security, retraining and reskilling of workers, wages, and working conditions.

The purpose of these agreements is to ensure that the introduction of automation does not lead to job losses or reduced compensation for workers. They also aim to ensure that workers are able to adapt to new technologies and are not left behind in the changing economy.

In some cases, these agreements may include provisions for the establishment of joint labour-management committees to oversee the introduction of automation and to monitor its impact on the workforce. These committees may be responsible for identifying areas where retraining or reskilling is needed, as well as for developing plans to mitigate the impact of automation on the workforce.

Overall, trade union automation agreements are seen as an important tool for protecting workers' rights and ensuring that the benefits of automation are shared fairly between workers and employers.

The relationship between trade unions and automation is complex and multifaceted, with both opportunities and challenges for organized labor. While automation has the potential to reduce jobs and bargaining power for workers, it can also create new opportunities for unionization and collective bargaining.

One potential future for trade unions in the face of automation is to actively engage with employers and policymakers to negotiate agreements that protect workers' rights and mitigate the negative impacts of automation. This could include measures such as retraining programs, job security provisions, and fair compensation for workers affected by automation.

Another potential future for trade unions is to focus on organizing workers in industries that are likely to be resistant to automation, such as healthcare, education, and hospitality. By building strong unions in these industries, workers can maintain their bargaining power and protect their rights in the face of technological change.

However, there are also challenges facing trade unions in the context of automation. As automation increases productivity and reduces the need for manual labor, it may be harder for unions to organize workers in some industries. In addition, there is a risk that unions could be left behind if they fail to adapt to new technologies and find ways to use automation to their advantage.

To remain relevant and effective, trade unions will need to be flexible and innovative in responding to the challenges of automation. This could include developing new forms of collective bargaining, building strategic partnerships with employers, and investing in technology training and education for union members. Ultimately, the future of trade unions in the face of automation will depend on their ability to adapt and evolve to meet the changing needs of workers in a rapidly changing world.

Chapter 12: Conclusion and final thoughts

Throughout this book, we have examined the negative effects of automation and its growing prevalence in various industries. We have discussed how automation can replace human workers and lead to unemployment, create economic inequality, limit human creativity and problem-solving ability, and even lead to the over-reliance on technology. We have also analyzed the potential risks associated with automation, including the possibility of autonomous systems malfunctioning or being used for harmful purposes.

Additionally, we have discussed the ethical implications of automation, including issues such as privacy, data security, and bias. We have provided specific case studies and examples of industries or companies where automation has had negative effects, and analyzed the factors that led to these negative effects, including the role of government policies, corporate decisions, and technological developments.

Furthermore, we have examined current trends in automation and predicted their potential impact on various industries and sectors. We have also provided recommendations for how organizations and governments can prepare for and respond to these potential consequences, including the use of alternative approaches to automation and strategies for overcoming the negative effects of automation, such as retraining workers, implementing ethical guidelines, and investing in social programs and safety nets.

Overall, it is clear that automation has the potential to greatly benefit society, but it also has significant drawbacks that must be addressed. As we move forward, it is important to consider the ethical implications of automation and take proactive steps to mitigate its negative effects, while also working to harness its potential benefits for the betterment of society as a whole.

As we come to the end of this book on the negative effects of automation, it is important to reflect on what we have learned and consider the implications for society as a whole. We have seen that while automation can bring many benefits, including increased efficiency and productivity, it can also have significant negative effects on individuals, communities, and society as a whole.

One of the key concerns is the potential for widespread job displacement, which can lead to unemployment and economic inequality. As technology continues to advance and more jobs become automated, it is crucial that we develop strategies for retraining workers and creating new job opportunities in order to mitigate these effects.

Another concern is the potential loss of human creativity and problem-solving ability. While automation can take on many routine tasks, it is important that we continue to value and develop the unique skills and talents that humans possess, including our ability to think creatively and critically.

We have also seen that increased reliance on automation and technology can have implications for privacy, data security, and bias. It is important that organizations and individuals take responsibility for ensuring that automation is used ethically and in ways that do not harm individuals or society as a whole.

Looking to the future, it is clear that automation will continue to play an increasingly important role in our lives. However, we must be mindful of the potential negative effects and work to develop strategies that can help us overcome these challenges.

In the end, it is up to us as a society to determine how we want to integrate automation into our lives and ensure that it serves the greater good. By working together and embracing innovation in responsible and ethical ways, we can create a future in which technology and humanity can thrive together.

As we conclude this book, it is important to reflect on the potential negative effects of automation and their implications for society. The increasing prevalence of automation in various industries and sectors raises concerns about job displacement, economic inequality, and the loss of human creativity and problem-solving ability. Additionally, ethical issues such as privacy, data security, and bias associated with automation require serious consideration.

While there are potential benefits to automation, including increased efficiency and productivity, it is essential to recognize and address the challenges it poses. It is crucial that ongoing dialogue and research are conducted to fully understand the impact of automation on society.

Collaborative efforts between various stakeholders, including businesses, governments, trade unions, and academia, are necessary to ensure that automation is used in an ethical and responsible manner. It is essential to implement strategies for mitigating the negative effects of automation, such as retraining programs and social safety nets, and to prioritize investment in education and skills development.

Trade unions have an important role to play in advocating for workers' rights and protecting their interests in the face of increasing automation. By engaging in collective bargaining and negotiating automation agreements with employers, trade unions can ensure that the introduction of automation is done in a way that benefits both workers and businesses.

The future of trade unions in the face of automation remains uncertain. While some may argue that automation will lead to the decline of traditional labour unions, others believe that unions will remain relevant as a means of protecting workers' rights and negotiating fair treatment in the face of increasing automation.

In conclusion, the potential negative effects of automation require serious consideration and ongoing dialogue. Collaborative efforts between various stakeholders are necessary to ensure that automation is used ethically and responsibly, and strategies for mitigating its negative effects must be implemented. While the future of trade unions in the face of automation remains uncertain, their role in advocating for workers' rights and negotiating fair treatment remains essential. As society continues to grapple with the challenges posed by automation, it is essential to prioritize the well-being of workers and society as a whole.

www.ingramcontent.com/pod-product-compliance
Lightning Source LLC
Chambersburg PA
CBHW072238230526
45466CB00024B/2102